THINK Like a Conservation Leader

Saving Species Today,
because tomorrow is too late!

By C S Wurzberger, the Critter Saver

Want to use this THINK Like a Conservation Leader curriculum in your school, home-schooling group, summer camp, daycare center, library, church, girl & boy scouts, 4-h group, campground, zoo, aquarium, wildlife conservation center, or other?

Contact us at Office@AwesomeAnimalAcademy.org for bulk order quantities and discounts.

ISBN: 9798880015825

Awesome Animal Academy
C S Wurzberger, the Critter Saver
P. O. Box 343
Marlboro, VT 05344

AwesomeAnimals.org
AwesomeAnimalAcademy.org

The cover photo was taken at The Wilds in Cumberland, Ohio. It is one of the largest conservation centers in the world and covers over 10,000 acres. A must-see animal adventure for animal lovers of all ages!

"What we appreciate, we preserve.
What we value, we conserve.
What we are taught, we understand.
And when we understand, we can come together
to protect the earth and its animals."
~ C S Wurzberger,
the Critter Saver

TABLE OF CONTENTS

A TOOLKIT FOR EDUCATORS

A HANDBOOK & TEACHING TOOLS

LET'S GROW THE NEXT GENERATION OF CONSERVATION LEADERS!

» **DOWNLOAD TOOLKIT**

Within the Toolkit, you'll find:

- **An eBook copy of 'THINK LIKE A CONSERVATION LEADER.'** (*Student Edition*) With worksheets, interactive projects, and discussion questions to grow as a conservation leader. (Also, a link to Amazon to purchase printed copies.
- **An eBook copy of 'GROW YOUNG LEADERS LIKE A SILVERBACK GORILLA.'** (*Educator Edition*) With worksheets, discussion questions, and interactive projects that grow the next generation of conservation leaders.
- **An all-inclusive presentation slide deck highlighting key topics** and discussion questions for online and in-person sessions. (Formats: PowerPoint and a printable PDF).
- **A set of companion, on-demand videos that bring the core leadership** concepts to life and inspire the launching of real-life conservation projects, locally and globally.
- **A set of fun, easy-to-start real-life conservation projects** to get the students started down the path of Conservation Leadership.

DOWNLOAD YOUR TOOLKIT TODAY: AwesomeAnimalAcademy.org

This Handbook is dedicated to
the Dodo Bird.
May you inspire the next generation
of young people to launch and lead their
own Critter Saver Project™ so no more
animals go extinct!

A LETTER FROM CS THE CRITTER SAVER

Hello Conservation Leader,

Welcome to the wild world of saving species!

I'm C S Wurzberger and I'd like to share a story with you that profoundly set the course for my life and why I've written this Handbook for you.

It all started when I was 10 years old.

I was asked to select an animal and write a paper for my English class. As I was flipping through the pages and pictures of a book, I came across the Dodo bird, a funny-looking creature that caught my attention.

As I read on, I discovered that the Dodo bird went extinct in 1681 because of overhunting and the introduction of animals that preyed on its young.

Boy, was I an upset little girl! I thought, "This bird could have been saved if people had cared a little more."

I remember coming home from school to tell my parents about the tragic news, yet they didn't seem to understand why I was so upset. I stomped my feet and shouted, "But we need to care. This bird will never be seen on the Earth again!"

It was hard for me to understand why no one in my life seemed to care about my love for animals. I was a quiet, shy little girl, and I didn't have the confidence and courage to speak up.

Well, now I do!

I spend my time passionately speaking up for animals and mentoring young people like you who want to make a difference in our world!

I don't want to see any more animals going extinct!

Most importantly, I'm here to help you gain the confidence, courage, and resources you need to speak up and protect the animals you love!

To also become a Critter Saver Champion™ who shares your voice, concerns, and solutions with the world!

Here are a few things going on in our world today:
- Mammals, birds, and fish populations have declined by 58% between 1970 and 2023.
- Over 100,000 marine mammals die each year from plastic pollution in the ocean.
- Freshwater species have declined 81% in the last two decades.

It's time for you to become the Leader and Champion you were meant to be and help protect our people, planet, and its animals!

In the upcoming chapters of this Interactive Handbook, you'll:
- Explore the world of Conservation Leadership
- Discover how leadership skills can improve your life
- Engage in real-life projects to protect the animals you love.

Plus, what's really exciting about this Handbook is it gives you a world of information and a detailed direction to ensure you T.H.I.N.K. Like a Conservation Leader!

Enjoy your day and the awesome animals around you,
CS Wurzberger, the Critter Saver

Celebrating The Life Of The Dodo Bird!

The Dodo bird is one of the most famous extinct animals. They were grey in color, stood about 3 feet tall, and weighed between 22 - 40 pounds.

Have fun exploring and finding answers to the below questions hiding in the picture:

Questions:

What year did the Dodo bird go extinct?

The Dodo bird lived on an island located in the Indian Ocean. What is the name of the island?

Did the Dodo live on the ground or in the trees?

How many eggs did the girl Dodo bird lay every other year?

1681

It lived and nested on the ground

One

Mauritius

WHAT OTHER ANIMALS DO YOU KNOW OF THAT ARE EXTINCT?

Do some detective work. Make a list of as many as you can, and list the years they went extinct.

"WHAT WE APPRECIATE, WE PRESERVE.
WHAT WE VALUE, WE CONSERVE.
WHAT WE ARE TAUGHT, WE UNDERSTAND.
AND WHEN WE UNDERSTAND, WE CAN COME TOGETHER
TO PROTECT THE EARTH AND ITS ANIMALS."

— C S WURZBERGER, THE CRITTER SAVER

PART #1

EXPLORE THE WORLD OF CONSERVATION LEADERSHIP

- 1-1 Introduction to Conservation Leadership.
- 1-2 Check Out the Four Pillars of Conservation.
- 1-3 Enjoy the Benefits of Being a Conservation Leader.
- 1-4 Make a Ripple Effect in Conservation Leadership.
- Summary – Action Steps.
- Things I learned in this section.
- Notes.

EXPAND YOUR IMPACT!

Join the Critter Saver Champions. A conservation club where ambitious and enthusiastic young people from around the world come together to save species, help habitats, and grow leadership skills.

CritterSaverChampions.com

INTRODUCTION TO CONSERVATION LEADERSHIP!

CONSERVATION LEADERSHIP IS ABOUT INSPIRING CHANGE AND TAKING ACTIONS THAT POSITIVELY IMPACTS OUR NATURAL WORLD.

IT'S ABOUT BEING A ROLE MODEL, SHOWING OTHERS HOW SMALL, DAILY CHOICES CAN ADD UP TO A BIG DIFFERENCE FOR OUR PLANET.

Every Action Counts:

- Every action, no matter how small, can have an impact on the environment. This includes everyday choices like recycling, conserving water, and choosing not to litter.

- Every person's actions affect the world around them, and it's essential to take responsibility for those actions. It's all about leading and inspiring others to do good.

CHECK OUT THE FOUR
PILLARS OF CONSERVATION!

The Four Pillars of Conservation are the foundation of protecting our planet and all its inhabitants. These pillars highlight the importance of safeguarding animal species, preserving natural environments, protecting marine life, and promoting human well-being by maintaining the health of our planet.

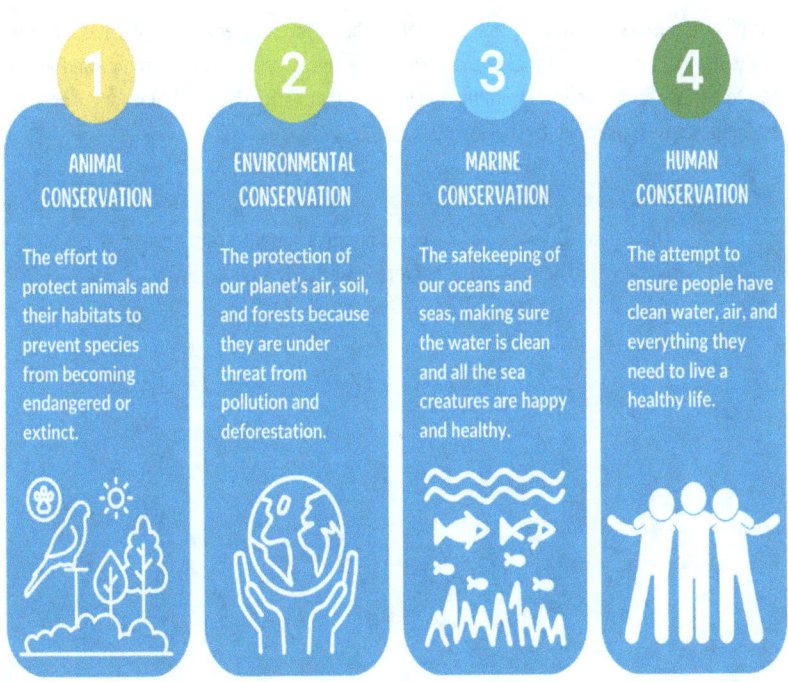

1

ANIMAL CONSERVATION

The effort to protect animals and their habitats to prevent species from becoming endangered or extinct.

2

ENVIRONMENTAL CONSERVATION

The protection of our planet's air, soil, and forests because they are under threat from pollution and deforestation.

3

MARINE CONSERVATION

The safekeeping of our oceans and seas, making sure the water is clean and all the sea creatures are happy and healthy.

4

HUMAN CONSERVATION

The attempt to ensure people have clean water, air, and everything they need to live a healthy life.

ALL DELIVERED WITH THE PURPOSE OF TEACHING PEOPLE TO CARE FOR OUR PLANET AND ALL ITS INHABITANTS!

CHECK OUT THE FOUR
PILLARS OF CONSERVATION!

Every pillar is connected, and by helping one segment, we often help the others. So, let's get creative, stay curious, and work together to make our world a healthier, happier place for all!

ANIMAL CONSERVATION:

This pillar focuses on protecting and preserving the different species that share our planet. It's like having a superhero for animals, making sure they have safe places to live and enough food to eat. By saving animals from threats like habitat destruction, pollution, and illegal poaching, we help maintain the balance of nature.

Discussion Question: If you could create a superhero to protect endangered animals, what powers would they have and why?

CHECK OUT THE FOUR PILLARS OF CONSERVATION!

Every pillar is connected, and by helping one segment, we often help the others. So, let's get creative, stay curious, and work together to make our world a healthier, happier place for all!

ENVIRONMENTAL CONSERVATION:

This pillar focuses on protecting the natural environments that are home to plants, animals, and even us! Environmental Conservation includes activities like planting trees, cleaning up litter, and using resources wisely to prevent pollution. It's like being Earth's gardener and caretaker, ensuring that our planet stays green and healthy.

2

Discussion Question: What simple daily actions can you take to help conserve the earth's resources around your home or school?

CHECK OUT THE FOUR
PILLARS OF CONSERVATION!

Every pillar is connected, and by helping one segment, we often help the others. So, let's get creative, stay curious, and work together to make our world a healthier, happier place for all!

MARINE CONSERVATION:

This pillar focuses on protecting the oceans, seas, and all the amazing life within them. From colorful coral reefs to mysterious deep-sea creatures, marine conservationists work to protect these underwater worlds from threats like overfishing, pollution, and climate change. It's about keeping the oceans healthy so they can continue to support life on Earth.

3

Discussion Question: Why do you think healthy oceans are important for life on land, and how can we help protect them?

CHECK OUT THE FOUR PILLARS OF CONSERVATION!

Every pillar is connected, and by helping one segment, we often help the others. So, let's get creative, stay curious, and work together to make our world a healthier, happier place for all!

HUMAN CONSERVATION:

This pillar focuses on Human Conservation, and it might sound a bit strange, but it's actually about ensuring that people can live sustainably with nature. This involves learning how to use natural resources wisely, promoting green technologies, and ensuring everyone can access clean air, water, and healthy environments. It's about taking care of each other and making sure future generations can enjoy our beautiful planet.

4 **Discussion Question:** How can you and your friends work together to promote human conservation and create a more sustainable future?

CHECK OUT THE FOUR
PILLARS OF CONSERVATION!

Every pillar is connected, and by helping one segment, we often help the others. So, let's get creative, stay curious, and work together to make our world a healthier, happier place for all!

WRAPPING UP:

Conservation is a team effort, and every action, no matter how small, makes a difference. Whether it's choosing to recycle, saving water, participating in a beach clean-up, or simply learning more about the natural world, we all have the power to contribute to the Four Pillars of Conservation. So, what will you do today to help protect our planet?

MAKE A LIST OF WHAT YOU CAN DO TODAY TO HELP:

ENJOY THE BENEFITS OF BEING A CONSERVATION LEADER!

Conservation Leadership will benefit your life by providing:

Growth and Fulfillment: As a leader, you'll grow leadership skills, expand entrepreneurial thinking, gain confidence, and much more. You'll also experience a deep sense of fulfillment from making a positive impact with your life.

Healthier Lifestyle: Engaging in conservation often leads to a healthier lifestyle. Whether it's spending more time outdoors or adopting eco-friendly habits, these changes can improve your physical and mental well-being.

Career Opportunities: Conservation Leadership can open doors to new educational opportunities and career paths in environmental science, sustainable development, and more.

How do you think being a Conservation Leader will benefit your life?

ENJOY THE BENEFITS OF BEING A CONSERVATION LEADER!

Part 1: Growth and Fulfillment

Write a brief paragraph describing a conservation project you want to lead. Be sure to explain how leading this project could help you grow your leadership skills and provide a sense of fulfillment.

Part 2: Healthier Lifestyle

List three eco-friendly habits you can adopt to lead a healthier lifestyle. Next to each habit, write a sentence explaining how it contributes to your physical or mental well-being.

1. _____: _____

2. _____: _____

3. _____: _____

Part 3: Career Opportunities

Research a career related to conservation. Write the name of the career, what they do, and how being a conservation leader could be beneficial in this role.

MAKE A RIPPLE EFFECT
IN CONSERVATION LEADERSHIP!

Becoming a Conservation Leader is more than just a title—it's a transformative journey that benefits you, your community, and the entire world. Like a pebble tossed into a pond, your actions and influence have a ripple effect, reaching far beyond your immediate surroundings.

© 2024 AwesomeAnimalAcademy.org

SEE HOW SYDNEY KICKED OFF A RIPPLE EFFECT!

A young person named Sydney plants milkweed in her backyard to support monarch butterflies. Here's how her small action triggers a ripple effect:

1

Initial Action: Sydney learns about the declining monarch butterfly population due to habitat loss and decides to plant milkweed, which is essential for their survival. Milkweed is the only plant the female Monarch will lay her eggs on and the only food source for the newly hatched caterpillars. Sydney starts with a few plants in her own backyard.

2

Peer Engagement: Sydney shares her project and its importance with friends at school. Inspired by her passion, several classmates began planting milkweed in their yards, also expanding the habitat for monarchs in the community.

3

Educational Campaign: Sydney is encouraged by all the support. She creates an educational campaign, including presentations for local schools and community groups, highlighting the plight of the monarchs and how simple actions like planting milkweed can aid their conservation.

SEE HOW SYDNEY
KICKED OFF A RIPPLE EFFECT!

A young person named Sydney plants milkweed in her backyard to support monarch butterflies. Here's how her small action triggers a ripple effect:

4

Community Butterfly Gardens: Sydney's efforts spark interest in the community, leading to the creation of several butterfly gardens in public spaces, parks, and schools. All feature milkweed and other nectar-rich plants for the Monarchs.

5

Conservation Advocacy: Sydney's initiative gains attention from local media and conservation groups. She uses this platform to advocate for broader conservation efforts, influencing others to take action in their communities and beyond.

So here you can see the complete ripple effect. Through Sydney's initial act of planting a few milkweed plants, she not only creates direct support for monarch butterflies but also inspires a community-wide movement towards conservation, showcasing the power of individual actions in generating widespread environmental impact.

NOW IT'S YOUR TURN TO OUTLINE
THE RIPPLE EFFECT DO YOU WANT TO MAKE!

DIRECTIONS: Have fun dreaming about the kind of ripple effect you want to make. Write your story below.

For example, after reading about the extinction of the Dodo Bird at 10 years of age, I knew that I wanted to teach young people how to help protect animals and make a global impact with my ripple.

Title: _____

Initial Action:

_____

_____

_____

_____

_____

_____

_____

_____

NOW IT'S YOUR TURN TO OUTLINE THE RIPPLE EFFECT DO YOU WANT TO MAKE!

2 Peer Engagement:

_____

_____

_____

_____

_____

3 Educational Campaign:

_____

_____

_____

_____

_____

_____

NOW IT'S YOUR TURN TO OUTLINE
THE RIPPLE EFFECT DO YOU WANT TO MAKE!

4 Community Engagement:

_____
_____
_____
_____
_____

5 Conservation Advocacy:

_____
_____
_____
_____
_____
_____

Summary - Action Steps

PART #1: Explore the World of Conservation Leadership

Mark off each item on the checklist to ensure you are moving closer to growing as a Conservation Leader!

☐ I have checked out the 4 pillars of conservation.

☐ I have explored the benefits of being a conservation leader.

☐ I have explored how Sydney kicked off a ripple effect.

☐ I have identified the ripple effect I want to make.

Congratulations! Now you're ready to Level Up and THINK Like a Conservation Leader!

Things I learned in this section

_____

_____

_____

_____

_____

_____

_____

_____

_____

_____

_____

_____

_____

_____

_____

_____

_____

_____

_____

My Notes

IGNITE YOUR CRITTER SUPERPOWERS

Choose from handbooks filled with engaging hands-on activities and key lessons in leadership, mastering mindset, confidence building, stewardship, entrepreneurship, podcasting, fundraising, and other awesome topics!

Also available with a companion video series and live, instructor-led programs (in-person or online)

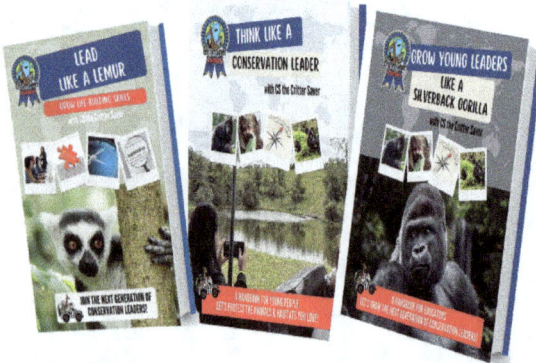

https://academy.awesomeanimals.org/ignite-critter-superpowers

LEVEL UP AND T.H.I.N.K. LIKE A TRUE CONSERVATION LEADER

The framework aims to develop skills in leadership, critical thinking, and problem-solving while fostering stewardship and empathy toward nature.

- 2-1 Level Up Your Leadership Skills.
- 2-2 THINK Like a Conservation Leader Framework.
- Summary – Action Steps.
- Things I learned in this section.
- Notes.

EXPAND YOUR IMPACT!

Join the Critter Saver Champions. A conservation club where ambitious and enthusiastic young people from around the world come together to save species, help habitats, and grow leadership skills.

CritterSaverChampions.com

Join Other Conservation Leaders at an Upcoming Critter Saver Summit.

Stop sitting on the sidelines ...

as our planet'shealth deteriorates and the number of endangered animals continues to grow.

Sign up for this 1-day, live, interactive virtual summit and learn how even the smallest actions can create a ripple effect of positive change and discover how to make a real difference in the fight against climate change, species extinction, and habitat loss.

RESERVE YOUR FREE CRITTER SAVER SUMMIT TICKET TODAY!

AwesomeAnimalAcademy.org

LEVEL UP YOUR LEADERSHIP SKILLS WHILE...

In this section, you'll discover how to level up your leadership skills and engage in strategies aimed at preserving all plants, animals, and their environments. It starts with Wildlife conservation is the practice of protecting wild species and their habitats in order to maintain healthy populations and ecosystems.

Let's get started:

PROTECTING ENDANGERED SPECIES:

Working to prevent species from becoming endangered or extinct.

HELPING HABITAT PRESERVATION:

Conserving natural habitats like forests, wetlands, grasslands, and oceans is crucial to maintaining biodiversity.

RESTORING ECOSYSTEMS:

Revitalizing areas where natural habitats have been degraded or destroyed.

PROMOTING BIODIVERSITY:

Maintaining a wide variety of species within their natural environments, recognizing that biodiversity is essential for ecosystem health and resilience.

RESPONSIBLY USING RESOURCES:

Promoting practices that ensure the responsible use of natural resources and making sure that human activities do not harm wildlife populations or their habitats.

RESEARCHING AND EDUCATING:

Understanding wildlife needs, threats, and conservation strategies is critical for their well-being. Also, education and public awareness campaigns.

HAVE FUN FILLING IN THE FOLLOWING WORKSHEETS!

LEVEL UP YOUR LEADERSHIP SKILLS WHILE PROTECTING ENDANGERED SPECIES

Wildlife conservation is the practice of protecting wild species and their habitats in order to maintain healthy populations and ecosystems. It involves a variety of efforts and strategies aimed at preserving all plants, animals, and their environments.

THE 6 MAIN OBJECTIVES OF WILDLIFE CONSERVATION INCLUDE:

1. Protecting Endangered Species: Working to prevent species from becoming endangered or extinct. This involves protecting their natural habitats, regulating hunting and fishing, and combating poaching and illegal wildlife trade.

DISCUSSION QUESTIONS:

Why do you think it's important to protect species from becoming endangered or extinct?

LEVEL UP YOUR LEADERSHIP SKILLS WHILE PROTECTING ENDANGERED SPECIES

DISCUSSION QUESTIONS:

How do human activities contribute to this issue, and what actions can we take to mitigate these effects?

2 LEVEL UP YOUR LEADERSHIP SKILLS WHILE HELPING HABITAT PRESERVATION

2. Helping Habitat Preservation: Conserving natural habitats like forests, wetlands, grasslands, and oceans is crucial to maintaining biodiversity. This includes setting up protected areas such as national parks and wildlife reserves where ecosystems can function naturally without human interference.

DISCUSSION QUESTIONS:

Discuss how the loss of natural habitats like forests and wetlands impacts not only the animals that live there, but also humans and the environment as a whole.

What role can protected areas such as national parks play in conservation efforts?

3

LEVEL UP YOUR LEADERSHIP SKILLS WHILE RESTORING ECOSYSTEMS

3. Restoring Ecosystems: Revitalizing areas where natural habitats have been degraded or destroyed. This includes reforestation, wetland restoration, and the reintroduction of native species.

DISCUSSION QUESTIONS:

Why is restoring degraded ecosystems important for wildlife conservation?

Can you think of an example where ecosystem restoration has successfully brought back a balanced natural environment?

4

LEVEL UP YOUR LEADERSHIP SKILLS WHILE PROMOTING BIODIVERSITY

4. Promoting Biodiversity: Maintaining a wide variety of species within their natural environments, recognizing that biodiversity is essential for ecosystem health and resilience.

DISCUSSION QUESTIONS:

How does biodiversity contribute to the health and resilience of an ecosystem?

What might happen to an ecosystem if its biodiversity significantly decreases?

5 LEVEL UP YOUR LEADERSHIP SKILLS WHILE RESPONSIBLY USING RESOURCES

5. Responsibly Using Resources: Promoting practices that ensure the responsible use of natural resources and making sure that human activities do not harm wildlife populations or their habitats.

DISCUSSION QUESTIONS:

What does 'sustainable use of natural resources' mean, and how can it help in conserving wildlife?

Discuss some practices that could be considered sustainable and how they benefit both humans and wildlife.

© 2024 AwesomeAnimalAcademy.org

6

LEVEL UP YOUR LEADERSHIP SKILLS WHILE RESEARCHING AND EDUCATING

6. Researching and Educating: Understanding wildlife needs, threats, and conservation strategies is critical for their well-being. Also, education and public awareness campaigns help to foster a sense of responsibility towards wildlife and encourage people to care more.

DISCUSSION QUESTIONS:

How can research and education play a role in wildlife conservation?

Share your thoughts on how increasing public awareness and knowledge about wildlife can contribute to conservation efforts.

T.H.I.N.K. LIKE A CONSERVATION LEADER FRAMEWORK

 T - TAKE the Critter Saver Path with all the right tools, resources, podcasts, mentors, coaches, global zoo directory, and your new super tool, AI. Enhance observation skills through nature exploration and storytelling.

 H - HELP Protect Homes and Habitat for All. Develop empathy and understanding of ecological balance through interactive habitat-related activities.

 I - INVESTIGATE Nature and the Wildlife Issues near you and around the world: Build research and critical thinking skills by investigating local wildlife challenges.

 N - NAVIGATE Challenges with Determination and Confidence: Focus on problem-solving and resilience through simulations and teamwork.

 K - KICK OFF with a 3-Step Critter Saver Project: Encourages leadership and planning, culminating in a real-world conservation project.

The framework aims to develop skills in leadership, critical thinking, and problem-solving while fostering stewardship and empathy toward nature. It offers participants a chance to make real-world impacts in conservation, providing a platform for community building and collaboration among young conservationists.

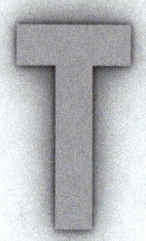

THINK LIKE A CONSERVATION LEADER

T - TAKE the Critter Saver Path with all the right tools, resources, podcasts, mentors, coaches, global zoo directory, and your new super tool, AI. Enhance observation skills through nature exploration and storytelling.

DISCUSSION QUESTIONS:

What are some everyday actions you can take to protect our environment and wildlife?

List the resources you like to use:

List the mentors who can help you:

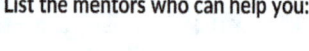

THINK LIKE A
CONSERVATION LEADER

H - HELP Protect Homes and Habitat for All. Develop empathy and understanding of ecological balance through interactive habitat-related activities.

DISCUSSION QUESTIONS:

Why is it important to protect wildlife and natural habitats?

Can you think of an animal that needs our help to survive? What can we do to help?

Do you think plants are as important to protect as animals? Why or why not.

THINK LIKE A
CONSERVATION LEADER

I - INVESTIGATE Nature and the Wildlife Issues near you and around the world: Build research and critical thinking skills by investigating local wildlife challenges.

DISCUSSION QUESTIONS:

> How do you think climate change affects wildlife? Can you give an example?

> Why do you think some animals become endangered or extinct?

> If animals could talk, what do you think they would say about the way humans treat the environment?

THINK LIKE A
CONSERVATION LEADER

N - NAVIGATE Challenges with Determination and Confidence: Focus on problem-solving and resilience through simulations and teamwork.

DISCUSSION QUESTIONS:

Identify a local conservation issue and brainstorm potential solutions. List option A below:

Identify a local conservation issue and brainstorm potential solutions. List option B below:

THINK LIKE A CONSERVATION LEADER

K - KICK OFF with a 3-Step Critter Saver Project: Encourages leadership and planning, culminating in a real-world conservation project.

The framework aims to develop skills in leadership, critical thinking, and problem-solving while fostering stewardship and empathy toward nature. It offers participants a chance to make real-world impacts in conservation by providing a platform for community building and collaboration among young conservationists.

First you can select some of our suggested conservation projects and then you can download your Critter Saver Handbook and launch some of your own custom projects.

HAVE A BLAST PROTECTING THE ANIMALS YOU LOVE AND THE HABITATS THEY CALL HOME!

DISCUSSION QUESTION:

If you were a conservation leader for a day, what would be the first action you'd take to help wildlife?

Summary - Action Steps

PART #2: LEVEL UP AND T.H.I.N.K. LIKE A TRUE CONSERVATION LEADER

Mark off each item on the checklist to ensure you are moving closer to growing as a Conservation Leader!

☐ I have leveled up my leadership skills.

☐ I have answered all the discussion questions.

☐ I have started down the Critter Saver Path.

☐ I have followed the T.H.I.N.K. Like a Conservation Leader Framework.

☐ I have started planning a Critter Saver Project™ to launch.

Congratulations! Now you're ready to move on to STEP 3: Launch Your Critter Saver Projects™.

Things I learned in this section

_____

_____

_____

_____

_____

_____

_____

_____

_____

_____

_____

_____

_____

_____

_____

My Notes

"UNLESS SOMEONE LIKE YOU CARES A WHOLE AWFUL LOT,
NOTHING IS GOING TO GET BETTER. IT'S NOT."

— THE LORAX - DR. SUESS

LAUNCH YOUR CRITTER SAVER PROJECTS™

- 3-1 Adopt the Role of a Conservation Leader.
- 3-2 Engage in Pre-Designed Projects.
- 3-3 Launch Your Own Critter Saver Projects™.
- Summary – Action Steps.
- Things I learned in this section.
- Notes.

Then, as a special bonus, I'll help you dream up, design, and launch your own real-life conservation projects. Turn your love for animals into real-world actions and adventures.

Engage in the mission to save bees to butterflies, wildlife to sea life, and exotic to endangered!

ADOPT THE ROLE OF A CONSERVATION LEADER!

A Conservation Leader handles the process of guiding, influencing, and leading efforts towards the protection, preservation, and sustainable management of natural resources and environments.

The role involves:

MAKING BIG PLANS:

Thinking of ways to help animals, plants, and places in nature stay safe and healthy for a long time.

PARTERNING WITH PEOPLE:

Teaming up with a variety of people, like friends, classmates, scientists, and other folks that care about protecting animals and thier habitats.

ADAPTING AND ADJUSTING:

Using cool new ideas and strategies when needed to take care of nature in the best way possible.

RAISING PUBLIC AWARENESS:

Teaching people why nature is important and what they can do to help protect it while sharing environmental issues and advocating for practices that support conservation.

USING RESOURCES WISELY:

Making sure that money, people, and tools are used in the best way to help nature and achieve the best conservation outcomes.

EDUCATING OTHERS:

Showing people how to care for nature and encouraging them to participate in conservation efforts and to adopt sustainable practices.

FILL IN THE FOLLOWING WORKSHEETS AND ENJOY YOUR NEW ROLE OF A CONSERVATION LEADER!

© 2024 AwesomeAnimalAcademy.org

ADOPT THE ROLE OF A CONSERVATION LEADER!

FOLLOW THE STEPS BELOW:

1. Making Big Plans: Thinking of ways to help animals, plants, and places in nature stay safe and healthy for a long time.

What big plans would you like to kick off?

ADOPT THE ROLE OF A CONSERVATION LEADER!

FOLLOW THE STEPS BELOW:

4. Raising Public Awareness: Teaching people why nature is important and what they can do to help protect it. Sharing environmental issues and advocating for policies and practices that support conservation.

What tactics will you use to get peoples attention? (Videos, posters, podcasts, etc.)

ADOPT THE ROLE OF A CONSERVATION LEADER!

FOLLOW THE STEPS BELOW:

5. Using Resources Wisely: Making sure that money, people, and tools are used in the best way to help nature and achieve the best conservation outcomes.

6. Educating and Empowering Others: Showing people how to care for nature and encouraging them to participate in conservation efforts and to adopt sustainable practices.

What resources would you like to line up?

How can you inspire others to engage in conservation?

ENJOY YOUR NEW ROLE OF A CONSERVATION LEADER!

ENGAGE IN PRE-DESIGNED PROJECTS

Here is your chance to help protect the animals you love while also growing your leadership skills, building confidence, expanding your entrepreneurial thinking, and protecting the animals you love!

Start with a pre-designed project:

PROJECT #1: Backyard or Local Park Biodiversity Study

PROJECT #2: Birdwatching in Your Backyard

PROJECT #3: Nature Journaling

PROJECT #4: Visit Local Zoos, Aquariums, Etc.

PROJECT #5: Endangered Species Research

Then, as a special bonus, I'll help you dream up, design, and launch your own real-life conservation projects. Turn your love for animals into real-world actions and adventures.

Engage in the mission to save bees to butterflies, wildlife to sea life, and exotic to endangered!

KICK OFF A CRITTER SAVER PROJECT™
BACKYARD OR LOCAL PARK BIODIVERSITY SURVEY

PROJECT #1

DIRECTIONS: Identify and count the number of different species of plants, insects, birds, and other animals you see in your backyard, local park or online webcam.

Use the worksheet on the next page. In the meantime, have fun drawing a picture of the things you see.

Skills Developed:
- Observation
- Species identification
- Data collection and recording

DRAW WHAT YOU SEE!

KICK OFF A CRITTER SAVER PROJECT™
BACKYARD OR LOCAL PARK BIODIVERSITY SURVEY

SURVEY

DIRECTIONS: Identify and count the number of different species of plants, insects, birds, and other animals you see in your backyard, local park or online webcam.

RECORD THE TYPES OF SPECIES YOU SEE	RECORD HOW MANY OF EACH SPECIES YOU SEE	RECORD WHERE YOU ARE WATCHING THEM

KICK OFF A CRITTER SAVER PROJECT™
BIRDWATCHING IN YOUR BACKYARD

DIRECTIONS: Head our birdwatching! Have fun identify and counting the number of different species of birds in their backyard or a local park.

Skills Developed:
- Patience
- Attention to detail
- Species identification

USE THE MAP BELOW TO COLOR IN WHERE YOU ARE BIRDWATCHING!

BIRDWATCHING LOG

This interactive log is designed to enrich your birding experiences. Record each bird sighting with key details like date, time, location, and species, and add a personal touch with your own photos and notes.

Sighting Details

Date: _____   Time: _____   Weather: _____

Location: _____

Bird Viewing Details

Bird Name: _____

Colors: _____

Size: _____

Beak Shape: _____

Special Markings: _____

Picture of Bird

Additional Notes:

_____

_____

_____

_____

_____

Picture of Bird

BIRDWATCHING LOG

This interactive log is designed to enrich your birding experiences. Record each bird sighting with key details like date, time, location, and species, and add a personal touch with your own photos and notes.

Sighting Details

Date: _____   Time: _____   Weather: _____

Location: _____

Bird Viewing Details

Picture of Bird

Bird Name: _____

Colors: _____

Size: _____

Beak Shape: _____

Special Markings: _____

Additional Notes:

_____

_____

_____

_____

_____

Picture of Bird

PLAN HOW YOU CAN HELP BIRDS

As you are observing the birds, keep an eye out to see if there are any issues impacting their livelihood. Do they need better access to water? Do they have enough food? Are people spraying pesticides near them?

Make a list of all the ways you can help them. Then, put a star next to the activity you would like to kick-off first.

☐ _____

☐ _____

☐ _____

☐ _____

☐ _____

☐ _____

☐ _____

☐ _____

☐ _____

☐ _____

☐ _____

☐ _____

☐ _____

PROJECT #3

KICK OFF A CRITTER SAVER PROJECT™
NATURE JOURNALING

DIRECTIONS: Have fun drawing or writing about your observations in nature. Also, make note of any environmental issues happening.

Skills Developed:
- Reflection
- Observation
- Artistic expression

Title: _____

KICK OFF A CRITTER SAVER PROJECT™
VISIT LOCAL ZOOS, AQUARIUMS, ETC.

PROJECT #4

DIRECTIONS: Learn firsthand about the efforts to protect, care for, and launch local & global conservation projects. Have fun visiting your local zoo, aquarium, nature center, and other places awesome animals life. You can visit in person or via webcams in the comfort of your home or classroom.

Skills Developed:
- Empathy for wildlife
- Understanding of conservation challenges
- Learning from experts

Start by making a dream list of all the animals you want to see and where you want to observe them. Anything is possible! You can see them in the wild, in captivity and even watch them with online animal cams.

SPECIES TO OBSERVE	ADDRESS TO VISIT	WILD - CAPTIVITY- ONLINE
PENGUINS	GEORGIAAQUARIUM.ORG	☆ ☆ ★
1		☆ ☆ ☆
2		☆ ☆ ☆
3		☆ ☆ ☆
4		☆ ☆ ☆
5		☆ ☆ ☆

Now use the worksheets on the following page to make your dream animal adventures come true!

DISCOVER WHERE YOUR FAVORITE ANIMALS LIVE!

These pages are designed for you to get out and explore nature and have a fun place to record all your Awesome Animal Adventures. There are lots of amazing places to visit animals in person, both in the wild and in captivity. You can explore your backyard, local park, zoo, aquarium, nature center, and more.

Also, depending on your lifestyle and location, many adventures can be enjoyed online from the comfort of your home, couch or classroom.

It is not necessary to take a long car ride to meet your favorite animals or even hop on a plane to head to the other side of the world. You can simply search the Internet for zoos, aquariums, nature centers, and other places that share streaming webcam feeds showcasing the animals.

Here is some great news! To make your animal search even easier, we have created an online directory for you to connect with all the top-rated animal-related organizations from around the world.

Visit **AwesomeAnimals.org** and search the most comprehensive online directory of zoos, aquariums, animal sanctuaries, wildlife parks, and conservation organizations available on the Internet. Use this directory for your research projects, homework, vacation planning, job/intern searches, and more!

AwesomeAnimals.org

VISIT AWESOME ANIMALS ONLINE OR IN PERSON

Here is a resource we have developed so you can easily find quality zoos, aquariums, nature centers and other places animals live around the world.

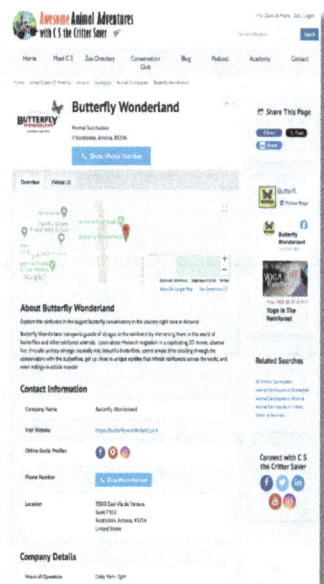

SCAN ME

With adult permission you can also search our online zoo directory:
AwesomeAnimals.org

AwesomeAnimals.org

NOTES FROM THE FIELD

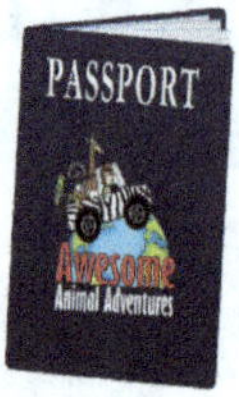

PASSPORT
Awesome
Animal Adventures

Record what you discovered about the local & global conservation projects you researched:

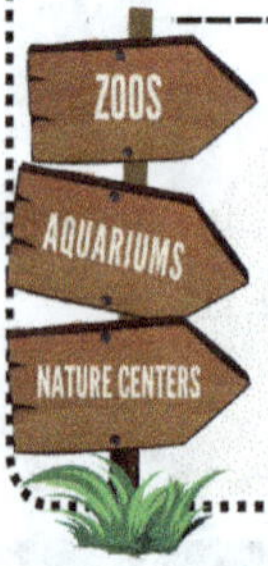

ZOOS
AQUARIUMS
NATURE CENTERS

KICK OFF A CRITTER SAVER PROJECT™
ENDANGERED SPECIES RESEARCH PROJECT

PROJECT #5

DIRECTIONS: Choose an endangered species, research it, and present findings through posters, presentations, or stories.

Skills Developed:
- Research
- Critical thinking
- Presentation skills

Pick an animal and describe what it looks like:

Diet: What do they eat & drink?

Issues: Why are they endangered?

Solutions: What can you do to help protect them?

CUSTOM PROJECTS

NOW IT'S YOUR TURN!

LAUNCH YOUR OWN CRITTER SAVER PROJECTS™

Launch real-life conservation projects that turn your love for animals into real-world actions and adventures.

Pick a category, choose your project, download your handbook, create a project, and make a true impact!

1
SAVE A SPECIES!

Launch real-life projects that protect the animals you love! Craft solutions and save species one project at a time!

2
HELP A HABITAT!

Restore natural homes, explore ecosystems, identify issues, and take action to repair one habitat at a time.

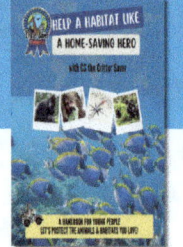

3
BUILD A BUSINESS!

Lay a foundation for your financial future and turn your ideas into a real-life, animal-related business.

Discover More At: AwesomeAnimalAcademy.org

DESIGN YOUR OWN CRITTER SAVER PROJECTS™ TO LAUNCH AND LEAD!

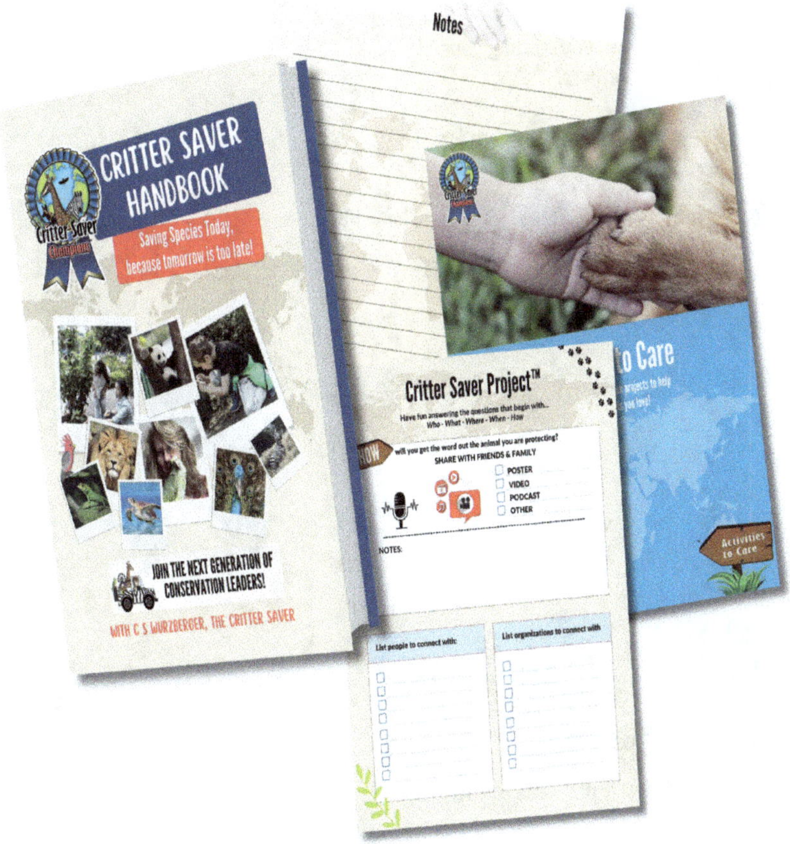

AwesomeAnimalAcademy.org

3-3

Critter Saver Project™

Have fun answering the questions that begin with...
Who - What - Where - When - How

WHAT animal are you helping?

WHAT supplies do you need?

- ☐
- ☐
- ☐
- ☐

WHEN are you starting your project?

WHAT type of project are you launching?

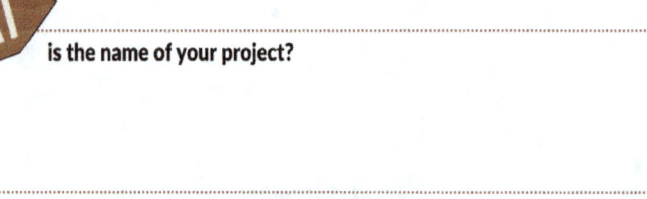

WHAT is the name of your project?

AwesomeAnimals.org

3-3

Critter Saver Project™

Have fun answering the questions that begin with...
Who - What - Where - When - How

WHERE is your project happening?

Example: park, zoo, forest, your house, etc.

WHO are your team members?

- ☐
- ☐
- ☐
- ☐

HOW much funding do you need? (If any?)

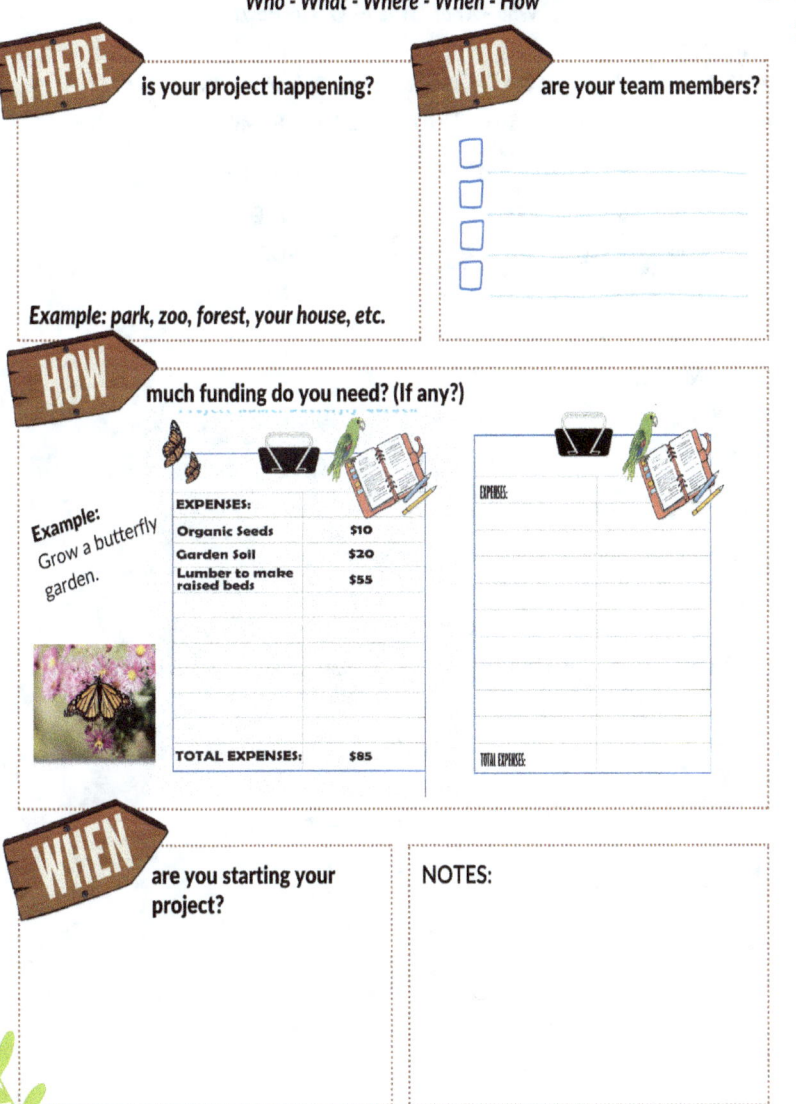

Example:
Grow a butterfly garden.

EXPENSES:

Organic Seeds	$10
Garden Soil	$20
Lumber to make raised beds	$55

| TOTAL EXPENSES: | $85 |

EXPENSES:

TOTAL EXPENSES:

WHEN are you starting your project?

NOTES:

AwesomeAnimals.org

3-3

Critter Saver Project™

Have fun answering the questions that begin with...
Who - What - Where - When - How

HOW will you get the word out about the animal you are protecting?

SHARE WITH FRIENDS & FAMILY

☐ POSTER
☐ VIDEO
☐ PODCAST
☐ OTHER

NOTES:

List people to connect with:	List organizations to connect with:
☐	☐
☐	☐
☐	☐
☐	☐
☐	☐
☐	☐
☐	☐
☐	☐

AwesomeAnimals.org

74

Additional Project Notes

Summary - Action Steps

PART #3: LAUNCH YOUR CRITTER SAVER PROJECTS™

Mark off each item on the checklist to ensure you are moving closer to growing as a Conservation Leader!

- [ ] I have adopted the role of a conservation leader.

- [ ] I have launched Project #1 - Backyard or Local Park Biodiversity Survey.

- [ ] I have launched Project #2 - Birdwatching in Your Backyard.

- [ ] I have launched Project #3 - Nature Journaling.

- [ ] I have launched Project #4 - Visit Local Zoos, Aquariums, Etc.

- [ ] I have launched Project #5 - Endangered Species Research Project.

- [ ] I have launched my own Critter Saver Project™.

Congratulations! Now you're ready to move on to the BONUS section: Explore Careers in Conversation!

Things I learned in this section

_____

_____

_____

_____

_____

_____

_____

_____

_____

_____

_____

_____

_____

_____

_____

_____

My Notes

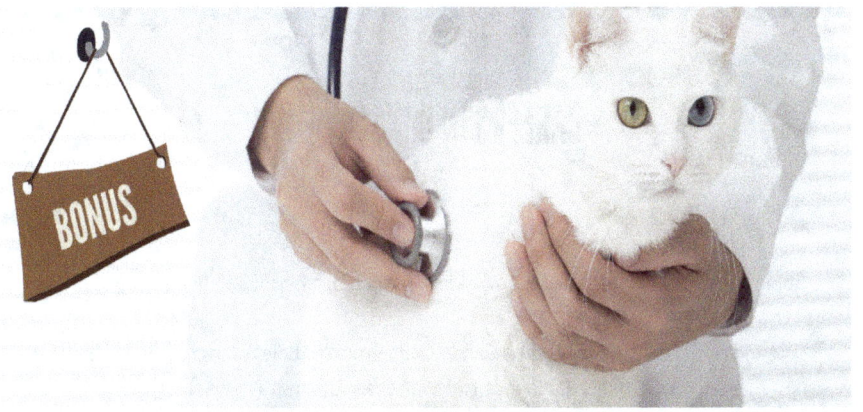

EXPLORE CAREERS IN CONSERVATION

Imagine having a job where you can protect animals, save forests, and help the earth breathe a little easier. That's what conservation careers are all about!

In this section, we'll dive into the world of those people who make it their life's work to look after our natural world. The wildlife conservationists who work to protect species, habitats, and ecosystems from extinction, damage, or disruption. To the scientists who study how to bring endangered animals back from the brink, you'll learn about all the cool jobs so you can choose one and make a difference also.

CAREERS IN CONSERVATION

Some of the roles that fall under a wildlife conservationist include:

1. Field Researcher: Conducts on-ground studies of species and habitats, collecting data that can inform conservation strategies. They might monitor species populations, behaviors, or health.

2. Wildlife Biologist: Specializes in studying animals and their behaviors. They may focus on one particular species or a group of species and their interactions within ecosystems.

3. Ecologist: Examines the relationships between organisms and their environment, looking at how changes in habitats impact species and the ecosystem's overall health.

4. Conservation Officer or Game Warden: Enforces laws related to wildlife, habitats, and natural resources. They may also be involved in educational outreach.

5. Wildlife Veterinarian: Provides medical care to wildlife, often in rehabilitation settings or sanctuaries. They might also study disease prevalence and transmission in wild populations.

CAREERS IN CONSERVATION

More roles that fall under a wildlife conservationist include:

6. Wildlife Educator: Raises awareness about wildlife and conservation issues through educational programs, workshops, and materials. This role might exist in zoos, nature centers, or schools.

7. Wildlife Photographer or Filmmaker: Uses visual mediums to tell stories about wildlife and conservation, raising awareness and drawing attention to critical issues.

8. Conservation Policy Analyst: Works on creating, analyzing, or advocating for policies and laws that support conservation efforts at local, national, or global levels.

9. Habitat Restoration Specialist: Focuses on restoring habitats that have been damaged or destroyed, ensuring they can support wildlife once more.

10. Wildlife Manager: Manages wildlife populations, sometimes in controlled settings like wildlife reserves, ensuring populations remain healthy and sustainable.

CAREERS IN CONSERVATION

More roles that fall under a wildlife conservationist include:

11. Zoo or Aquarium Conservationist: Works in captivity settings, focusing on breeding programs, education, and sometimes reintroduction efforts for endangered species.

12. Community Conservationist: Engages local communities in conservation efforts, recognizing the integral role communities play in the success of many conservation initiatives.

13. Conservation Fundraiser or Grant Writer: Secures funding for conservation projects through donations, grants, or other funding mechanisms.

14. Geographic Information Systems (GIS) Specialist: Uses spatial analysis tools to map habitats, track animal movements, or plan conservation strategies.

15. Marine Biologist or Marine Conservationist: Specializes in ocean and marine life conservation, addressing threats like overfishing, coral bleaching, or pollution.

These roles represent just a subset of the potential careers in wildlife conservation. Many conservationists also work in interdisciplinary teams, collaborating with professionals from other fields, such as anthropology, sociology, economics, or law, to achieve holistic conservation outcomes.

CAREERS IN CONSERVATION

DISCUSSION QUESTIONS:

If you are dreaming of a career working with animals, check out the Careers with Critters Masterclass!
AwesomeAnimals.org

What's one dream you have for your future?

IF YOU COULD CHOOSE ANY CAREER, NO MATTER HOW WILD, WHAT WOULD IT BE AND WHY?"

Let your imagination go wild! Have fun writing about your dream career!

Title: _____

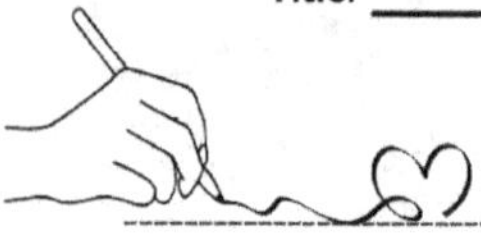

_____

_____

_____

_____

_____

_____

_____

_____

_____

_____

_____

DREAMS AND ASPIRATIONS

DISCUSSION QUESTIONS:

If you could have any superpower, what would it be and why?

If you could travel anywhere in the world and meet any animal, where would you go and who would you meet?

Summary - Action Steps
BONUS: CAREERS IN CONSERVATION

Mark off each item on the checklist to ensure you are moving closer to growing as a Conservation Leader!

- [ ] I have explored the different types of careers in conservation.

- [ ] I have thought about one dream for my future.

- [ ] I have chosen my dream career.

- [ ] I have chosen my superpower.

- [ ] I have chosen where I want to travel and which animal I want to meet.

Congratulations! Now you're ready to move on to your NEXT STEPS and IGNITE Your Critter Super Powers!

Things I learned in this section

_____

_____

_____

_____

_____

_____

_____

_____

_____

_____

_____

_____

_____

_____

_____

Notes

YOUR NEXT STEPS & RESOURCES!

As you turn the pages of this Handbook and reflect on all the conservation leadership skills you have learned, consider taking the next step on your path as a leader and become a caregiver for the animals you love and the habitats we all call home!

Join me, C S Wurzberger, the Critter Saver, as I guide you through the next series of steps so you can grow as a conservation leader!

It is your turn to join the true Critter Saver Champions™!

YOUR NEXT STEPS & RESOURCES!

Follow these 3 fun steps and join the true Critter Saver Champions™.

Each step will ensure you will grow leadership skills, expand entrepreneurial thinking, and build confidence, which will give you the tools, mindset, and added advantages in your life.

1

DOWNLOAD YOUR CRITTER SAVER HANDBOOK (FREE GIFT)

This interactive, project-based handbook will guide you through each step and give you fun, detailed directions to launch your own Critter Saver Project™. Have fun saving bees to butterflies, wildlife to sea life, and exotic to endangered!

https://academy.awesomeanimals.org/sign-up-page-critter-saver-handbook-gift

2

IGNITE YOUR CRITTER SUPERPOWERS

This handbook series gives you life-building topics: Lead Like a Lemur, Grow Confidence Like a Cheetah, Present Like a Parrot, and other awesome topics!

https://academy.awesomeanimals.org/ignite-critter-superpowers

3

JOIN THE CONSERVATION CLUB

Critter Saver Champions is a Conservation Club where young people like you can join virtually from anywhere in the world. It has on-demand videos, interactive learning labs, fun activities, and more. You also will launch projects that save species, help habitats, and facilitate fundraising plans to protect the animals you love.

https://academy.awesomeanimals.org/critter-saver-champions

AwesomeAnimalAcademy.org

1 DOWNLOAD YOUR
CRITTER SAVER HANDBOOK™

A FREE GIFT

Get ready to jumpstart your journey into wildlife conservation and design your own **Critter Saver Projects™.** Join other animal lovers from around the globe who are leading projects that are:

- Saving Monarchs by planting milkweed gardens,
- Eliminating the use of plastic bags to help save sea turtles,
- And so much more!

Have fun saving bees to butterflies, wildlife to sea life, and exotic to endangered!

https://academy.awesomeanimals.org/sign-up-page-critter-saver-handbook-gift

2 IGNITE YOUR CRITTER SUPERPOWERS

Choose from handbooks filled with engaging hands-on activities and key lessons in leadership, mastering mindset, confidence building, stewardship, entrepreneurship, podcasting, fundraising, and other awesome topics!

Also available with a companion video series and live, instructor-led programs (in-person or online)

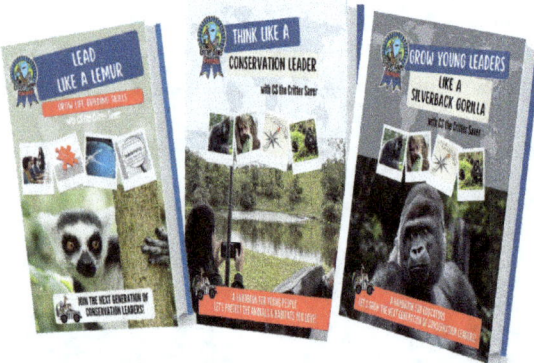

https://academy.awesomeanimals.org/ignite-critter-superpowers

JOIN THE CONSERVATION CLUB

Critter Saver Champions is a Conservation Club where young people like you can join virtually from anywhere in the world. It has on-demand videos, interactive learning labs, fun activities, and more. You also will launch projects that save species, help habitats, and facilitate fundraising plans to protect the animals you love.

Sneak Peek of the Club

‹ Back to Library

CRITTER SAVER **CLUB**

Start Here ⌄   Newest Posts   Challenges   Meetups

NEW IN THE LABS: Explore the benefits of learning leadership skills - Test Description   See announcement

Critter Saver Club   ...

Turn your love for animals into real-world actions and adventures! Attend THE LABS and learn life-building skills from the animal kingdom. Launch real-life protection projects IN THE FIELD. Visit THE LOUNGE and mingle with fellow animal lovers. Use your all-access pass IN THE LIBRARY to watch on-demand videos and download your handbooks.

👥 Find Members

☆ Leaderboard

Circles   🔍

CRITTER SAVER CHAMPIONS ⌄

📋 THE LABS
I'm inviting you to our Mem...

📋 IN THE FIELD
IN THE FIELD, you'll meet t...

📋 THE LOUNGE
IN THE LOUNGE, offers a n...

The Critter Saver Club is Made Up of 4 Sections:

1. The Labs

What are the Labs all about?

IN THE LABS, you'll become a junior scientist and conservation leader. Each session covers various topics like Lead Like a Lemur and Build a Business Like a Beaver. You'll grow leadership skills, build confidence, expand entrepreneurial thinking, and so much more. Meet awesome animals, uncover the issues that impact their well-being, and protect the animals you love.

2. In The Field

What happens in the field?

IN THE FIELD, you'll meet the trailblazers, the wildlife conservationists who work to protect species, habitats, and ecosystems from extinction, damage, or disruption. Imagine having a job where you can protect animals, save forests, and help the animals breathe a little easier. Now, you do. Join in the fun and engage in real-life protection projects during each IN THE FIELD session.

3. The Lounge

What's happening in the Lounge?

IN THE LOUNGE, offers a cozy digital space that allows you to mingle with fellow animal lovers, share details of your own conservation projects, seek answers to burning questions, share your thoughts and solutions on real-life topics, and keep up-to-date on what's new and exciting in the world of wildlife conservation.

Adventures to Connect

Activities to Care

Actions to Champion

https://academy.awesomeanimals.org/critter-saver-champions

EXPLORE FUN RESOURCES!

Turn your love for animals into real-world actions and adventures. Have fun exploring these additional resources:

VISIT THE GLOBAL ZOO DIRECTORY (ONLINE & IN PRINT)

Do you love traveling and meeting animals from around the world? This global directory will help you explore top-quality zoos, aquariums, nature centers, and more.

https://www.awesomeanimals.org/zoo-directory

MEET YOUR PERSONAL COACH CS THE CRITTER SAVER

C S Wurzberger specializes in working with ambitious and enthusiastic young people who want to launch their own conservation initiatives and socialpreneur projects that protect the animals they animals love !

https://www.awesomeanimals.org/cs_the_critter_saver

BRING A SPECIES SAVING PROGRAM TO YOUR...

School, homeschooling group, summer camp, library, Girl & Boy Scouts, 4-H group, campground, zoo, aquarium, wildlife conservation center, or anywhere else in the world.

https://www.awesomeanimals.org/cs_the_critter_saver

AwesomeAnimalAcademy.org

VISIT THE GLOBAL ZOO DIRECTORY

Search the online directory for quality zoos, aquariums, nature centers, animal sanctuaries, and other awesome places where animals live worldwide.

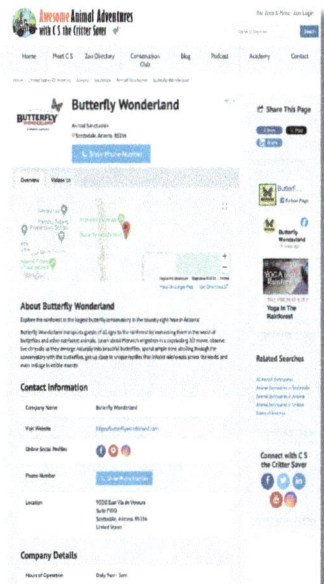

SCAN ME

With adult permission you can also search our online zoo directory:
AwesomeAnimals.org

AwesomeAnimals.org

LIST 10 PLACES
YOU DREAM OF VISITING!

What animals do you look forward to seeing in person?

1 _____

2 _____

3 _____

4 _____

5 _____

6 _____

7 _____

8 _____

9 _____

10 _____

ZOOS

AQUARIUMS

NATURE CENTERS

TRAVEL NOTES:

List any important travel information as you plan your trips to zoos, aquariums, and other places animals live!

PASSPORT

Awesome
Animal Adventures

ZOOS

AQUARIUMS

NATURE CENTERS

Meet C S Wurzberger, the Critter Saver

A certified high-performance coach who helps young people launch local & global conservation projects!

C S specializes in working with ambitious and enthusiastic young people who want to launch their own conservation initiatives and socialpreneur projects that protect the animals they animals love while sparking curiosity, building confidence and compassion, growing leadership skills, expanding entrepreneurial thinking, and developing respect and empathy for all!

In Memory of the Dodo Bird

C S BELIEVES:

1. **In the importance of protecting** all living creatures so they will be around for future generations to enjoy!
2. **That when people are able to** connect with animals they will, in turn, care more about their conservation!
3. **That connecting with animals** and nature can recharge our everyday lives plus improve our mental, physical, and spiritual health!

"What we appreciate, we preserve.
What we value, we conserve.
What we are taught, we understand.
And when we understand, we can come together
to protect the earth and its animals."
-- C S Wurzberger,
the Critter Saver

MORE ABOUT C S:

She is a Certified High-Performance Coach, Conservation Leader, Author, Podcaster, and Educator with 35+ years of experience, C S offers online and in-person programs at selected youth organizations, summer camps, schools, and homeschooling groups.

Plus, in partnership with zoos, aquariums, wildlife conservation centers, and farms, she provides the tools and knowledge to launch and lead a collection Critter Saver Projects™ around the world!

FORMER DIRECTOR:

C S is also the former director of a 150-acre, 300-animal, USDA-licensed educational petting zoo. There, she created, implemented, and promoted all of their educational programs. Plus, she experienced the joy of bottle-raising numerous animals such as Walter the Wallaby, Parachute the Field Mouse, Ted the Cheviot Lamb, Andy the Aoudad, and many others who touched her heart!

Bring a Species Saving Program to Your...

School, homeschooling group, summer camp, library, girl & boy scouts, 4-h group, campground, zoo, aquarium, wildlife conservation center, or any place in the world.

Together, We Can Grow the Next Generation of Conservation Leaders!

CS the Critter Saver travels around the U.S., presenting programs that inspire young people to learn key lessons in leadership, confidence building, stewardship, entrepreneurship, podcasting, and other awesome animal topics!

Let's discuss how the Critter Saver Champions™ curriculum can expand your education programs.

BOOK A CALL TODAY!
AwesomeAnimalAcademy.org

"ACT AS IF WHAT YOU DO MAKES A DIFFERENCE. IT DOES."

— WILLIAM JAMES

Turn your love for animals into real-world actions and adventures. Simply follow these 3 steps:

1

Join the Critter Saver Champions™!

Jumpstart your journey in wildlife conservation with sessions tailored to your passion and interests. This is your first step towards making a meaningful impact in the lives of the animals you love.
Learn leadership grow confidence, and expand your entreprenurial thinking.

2

Work with a conservation coach!

Collaborate with a certified high-performance coach who understands the nuances of wildlife conservation with a mix of leadership and social entrepreneurship.
Together, you'll clarify your goals and devise a plan that resonates with your mission to protect wildlife.

3

Make a lasting impact in your life and theirs!

Soar with a clear plan and newfound knowledge.
Watch as your conservation efforts start making a real difference. You're not just learning - you're growing confidence, expanding leadership skills, and taking actions that benefit the environment and the animals you care about.

AwesomeAnimalAcademy.org

Unleash Your Potential and Dive into Action for Wildlife!

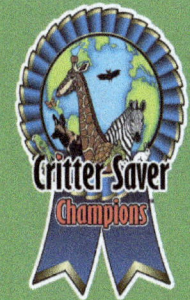

www.ingramcontent.com/pod-product-compliance
Lightning Source LLC
Chambersburg PA
CBHW070749290526
45795CB00002B/540

* 9 7 9 8 8 8 0 0 1 5 8 2 5 *